ANNUNCIAT

2 411323 002

945978

821TOML

TOMLINSON, C.

Annuniciations: Poems

1990   £5.95

This book is due for return on or before the
last date shown above. It may be renewed at
the Library or by post or telephone.

OXFORDSHIRE COUNTY LIBRARIES

Also by Charles Tomlinson:

Collected Poems (revised edition), 1987
The Return, 1987

ANNUNCIATIONS

Charles Tomlinson

Oxford   New York

OXFORD UNIVERSITY PRESS

1989

Oxford University Press, Walton Street, Oxford OX2 6DP
Oxford New York Toronto
Delhi Bombay Calcutta Madras Karachi
Petaling Jaya Singapore Hong Kong Tokyo
Nairobi Dar es Salaam Cape Town
Melbourne Auckland

and associated companies in
Berlin Ibadan

Oxford is a trade mark of Oxford University Press

First published as an
Oxford University Press paperback 1989

British Library Cataloguing in Publication Data

Tomlinson, Charles, 1927
Annunciations. — (Oxford poets)
I. Title II. Series
821'.914
ISBN 0–19–282680–8

Library of Congress Cataloging in Publication Data
Tomlinson, Charles, 1927–
Annunciations/Charles Tomlinson.
p.cm.
I. Title.
PR6039.O349A84 1989 821'.914 — dc20 89–31731
ISBN 0–19–282680–8

Typeset by Wyvern Typesetting Ltd.
Printed in Great Britain by
J.W. Arrowsmith Ltd., Bristol

To Brenda

ACKNOWLEDGEMENTS

Acknowledgements are due to the editors of the following:
Agenda, Antaeus, Culturas (Madrid), *Numbers, Paris Review, Partisan Review, Poetry* (Chicago), *Poetry Nation Review, Poetry Review, The Poetry Society Bulletin, Scripsi* (Australia), the *Times Literary Supplement, Vuelta* (Mexico City). 'Variation' was commissioned by the editor of *Poems for Shakespeare.* These poems appeared for the first time in *The Hudson Review*: 'The Butterflies', 'Far Point', 'The Cycle', 'Parking Lot', 'The Santa Fe Railroad', 'Chance', 'Letter to Uehata', 'Harvest', 'For a Godchild', 'A Ruskinian Fable Retold', 'Chronochromie', 'The Garden'. 'Oxen' appeared in *First and Always, Poems for Great Ormond Street Children's Hospital.*

CONTENTS

ANNUNCIATION

The cat took fright
at the flashing wing of sunlight
as the thing
entered the kitchen, angel of appearances,
and lingered there.

What was it the sun
had sent to say
by his messenger, this solvent ray,
that charged and changed
all it looked at, narrowing even the eye of a cat?

Utensils caught a shine
that could not be used, utility
unsaid by this invasion
from outer space, this gratuitous occasion
of unchaptered gospel.

'I shall return,' the appearance promised,
'I shall not wait for the last
day – every day
is fortunate even when you catch
my ray only as a gliding ghost.

What I foretell
is the unaccountable birth each time
my lord the light, a cat and you
share this domestic miracle:
it asks the name anew

of each thing named
when an earlier, shining dispensation
reached down into mist
and found the solidity
these windows and these walls surround,

and where each cup,
dish, hook and nail
now gathers and guards the sheen
drop by drop
still spilling-over
out of the grail of origin.'

THE BLADE

I looked to the west:
I saw it thrust
a single blade
between the shadows:
a lean stiletto-shard
tapering to its tip
yellowed along greensward,
lit on a roof that lay
mid-way across its path
and then outran it:
it was so keen,
it seemed to go
right through and cut
in two the land
it was lancing. Then
as I stood,
the shaft shifted,
fading across grass,
withdrew as visibly as the sand
down the throat of an hour-glass:
you could see time
trickle out, a grainy
lesion, and the green
filter back to fill
the crack in creation.

VARIATION

And there is nothing left remarkable
Beneath the visiting moon.
                Antony and Cleopatra, 4.vii

What is left remarkable beneath the visiting moon
   Is the way the horizon discovers itself to be
The frontier of a country unseen till this:
   Soon the light will focus the whole of it
Under one steadying beam, but now in rising
   Still has to clear the brow of a hill
To unroll the unmapped differences here,
   Where the floor of the valley refuses to appear
Uncoped by the shadow of its flank: it is the speed
   That accompanies this deed of climbing and revealing
Marks the ascent: you can measure out the pace
   Of the unpausing visitant between tree and tree,
Setting each trunk alight, then hurrying on
   To shine back down over the entire wood
It has ignited to flicker in white. Free
   From the obstructions it has come burning through
It has the whole of the night sky to review
   The world below it, seeming to slow
And even to dream its way. It does not arrive alone
   But carries the memory of that spread of space
And of the aeons across which it has shone till now
   From the beginning. This is the illumination it pours
Into the shadows and the watcher's mind,
   As it touches on planes of roofs it could not foresee
Shaping and sharing its light when it set out
   In a rain of disintegrating comets, of space creating.

IN NEW MEXICO

Where had the clouds come from? A half-hour since
   There was none, then one, the size
Of a man's hand that unclenched, spread, grew
   In no time into an archipelago on blue –
Blue that had begun with only an eagle in it
   And, even now, offered its unfilled miles
But not for long. The inter-breedings,
   The feathery proliferations of cloud-boats
Careless of anchorage, set out in flotilla
   Over this dry sea-bed they pied with their shadows
Until they covered the entire sea above.
   In the spaces between, you saw through
To an upper blue where whole webs
   Of thread billowed and floated free,
As if ready to be wound down and repair
   The least signs of thinning away there
Of what was already more than enough:
   Throughout the vagrant disorder of the sky
Excess was to be the order of the day.
   By afternoon, scarcely the rocks held their own:
The ground was no more than a screen
   Onto which the heavens could project themselves
And alter it all at will. True,
   You could pick up a stone and feel
That was a tight world still, but the white
   Seed of the cottonwood as the breeze that was shaping
      the cloud
Took hold of the tree and shook it,
   Was drifting across the sand like shreds of sky.

ABOVE THE RIO GRANDE

for Claude-Marie Senninger

The light, in its daylong play, refuses
   The mountains' certainty that they
Will never change – range on range of them
   In an illumination that looks like snow.
On this afternoon when the clouds are one impending grey
   Above the Rio Grande, the light will not obey
Either the clouds' or the rocks' command
   To keep its distance from them. It shifts and shows
Even the cloudshadows how to transform
   The very stones by opening over them
Dark wings that cradle and crease their solidity –
   As if to say: I gather up the rocks
Out of their world of things that are merely things,
   I call dark wings to be bearers of light
As they sail off the shapes they pall
   And, in their wake, leave this brightening snowfall
That melts and is renewed. Yet if the light
   Washes the rocks away, the rocks remain
To tell what it is, and only so
   Can they both flow and stay, and the mind
That floating thing, steady to know itself
   In all the exceedings of its certainty,
As here, beneath the expanses deepening
   Through the cloud-rock ranges of evening sky.

THE SANTA FE RAILROAD

The tri-toned whistle of the night train
   Starts up a blues that goes no further
Than this one repeated phrase, and yet it goes
   Far enough. The desert has no need
To declare itself by fanfare. It is itself indeed
   As you and I might be only in Eden.
And so the three tones – flattened on the final note –
   Float out over so much of space
There is no more to say except: these notes
   Express enough of possibility and of sadness, too,
To tell the extent and loneliness of the continent.
   Is that not so? – I ask Adelicia, who replies
When you catch the trainsounds here, it's going to rain.

PARKING LOT

Rain has filled all the streets
   With mirrors: the desert drank them;
The asphalt, mercury to their flickerings,
   Confronts the city with itself in fragments –
PARKING reading as ARK reversed
   (I took it for Russian). A slice of building
Has got in on the reflection of a car
   Disintegrating and reassembling between raindrops:
A piece of tree dances across the wrinkling image.
   Parked, I see half a man
Float by, and hear from the missing portion
   A throat being cleared and then
He walks away into his own entirety
   As I look up from the pool to where
He is passing the tree. It has pieced itself together now
   Beside a yellow fire hydrant
Into an appreciative maple, open-leaved
   To the wet spring morning, the yellow
Dwarf that accompanies it, the sole thing
   Small enough to be mirrored there whole.

AILANTHUS

'Ladders to heaven' is the Indian name
for these trees that in the heat
surreptitiously root themselves and grow
where nobody sees them –
as this one, live, lithe, grey,
out of the woodpile. Though its boughs
are supple and almost white,
it is not quite a willow.
Attacked, hacked back, it renews
and grows again up against the house
fingering the flank of the adobe wall
in each breath of wind.
'Rank' is the word
the poet chose for its odour,
but it is neither prized nor praised
for that. No need to climb
the ladder: the heaven
it leads to
lies downwards out of the sun
and under the cool liquidity of its shade.

CATALPA

A hill of leaves,
the tree three days ago
stood bare of bloom.
White torches flared
below the height they were to climb
and light from foot to crown
and did. It was the close of May
when roses overblown already
burned-on still,
but now another fire
struck up the tree:
more foam than flame,
spray after spray
riding the tide of wide and deep-green leaves,
this risen sea of flowers
had flooded every branch till all
the tree was one swaying festival.

ACOMA: THE ROUND-UP

I took the cliff-path down
– hand-holds and steps of rock –
and climbed to the plain
from the mesa top.

Singing (I thought it was)
wound on up the stair
in bursts of resonance
through the stony fissure.

It was the cattle
they were herding together,
closing the circle in,
rider by rider;

then driving them on
to await in the shade
of the cottonwood trees
those bringing the strays.

So the herdsmen sit
with scarcely a word,
diffusing a calm that
seems to flow through the herd.

And the only sound
is the metal whir
from the bladed wheel
that raises the water,

as the breeze spins it round
and cools all the air
under the leaves
where the gathered beasts are.

HERE AND THERE

There creeks had dried to saltstains in their beds,
To grains of the surrounding desert: veils
Of dust wove in a red transparency
Along the currents of the restless air,
Mile-wide chimeras changing dust to fire.
Now, more than mountains shut us from that scene
Our sun is westering towards, and yet we share
In a world imagined that is really there,
That reaches for our thoughts across the space
And time that free us, our imaginings
No more or less than links in that one chain,
Plains, peaks and seas, the tideline, estuaries
All stored within the mind's bright satellite
As it comes curving homeward. Here, we ride
On the muscularity and shift of tides
We can no longer see – the repetition
Of rippled dunes of sand, sustained and lit
Beyond our gaze, beyond this one day's lapse
Bright at the brink, and still suffusing night
With the waves on waves of its continuing fire.

THE PLAZA

People are the plot
and what they do here –
which is mostly sit
or walk through. The afternoon sun
brings out the hornets:
they dispute with no one, they too
are enjoying their ease
along the wet brink of the fountain,
imbibing peace and water
until a child arrives,
takes off his shoe
and proceeds methodically
to slaughter them. He has the face
and the ferocious concentration
of one of those Aztec gods
who must be fed on blood.
His mother drags him away, half-shod,
and then puts back the shoe
over a dusty sock.
Some feet go bare, some sandalled,
like these Indians who march through
– four of them – carrying a bed
as if they intended to sleep here.
Their progress is more brisk
than that of the ants at our feet
who are removing – some
by its feelers, some
supporting it on their backs –
a dead moth
as large as a bird.
As the shadows densen
in the gazebo-shaped bandstand
the band are beginning to congregate.
The air would be tropical
but for the breath of the sierra:
it grows opulent on the odour
of jacaranda and the turpentine
of the shoeshine boys
busy at ground-level,
the squeak of their rags on leather
like an angry, repeated bird-sound.

The conductor rises,
flicks his score with his baton –
moths are circling the bandstand light –
and sits down after each item.
The light falls onto the pancakes
of the flat military hats
that tilt and nod
as the musicians under them
converse with one another – then,
the tap of the baton. It must be
the presence of so many flowers enriches the brass:
tangos take on a tragic air,
but the opaque scent
makes the modulation into waltz-time seem
an invitation – not to the waltz merely –
but to the thought that there may be
the choice (at least for the hour)
of dying like Carmen
then rising like a flower.
A man goes by, carrying a fish
that is half his length
wrapped in a sheet of plastic
but nobody sees him. And nobody hears
the child in a torn dress
selling artificial flowers,
mouthing softly in English, 'Flowerrs'.
High heels, bare feet
around the tin cupola of the bandstand
patrol to the beat of the band:
this is the democracy
of the tierra templada – a contradiction
in a people who have inherited
so much punctilio, and yet
in all the to-and-fro
there is no frontier set:
the shopkeepers, the governor's sons,
the man who is selling balloons
in the shape of octopuses, bandannaed heads
above shawled and suckled children
keep common space
with a trio of deaf mutes
talking together in signs,
all drawn to the stir
of this rhythmic pulse

they cannot hear. The musicians
are packing away their instruments:
the strollers have not said out their say
and continue to process
under the centennial trees.
A moon has worked itself free
of the excluding boughs
above the square, and stands
unmistily mid-sky, a precisionist.
The ants must have devoured their prey by this.
As for the fish . . . three surly Oaxaqueños
are cutting and cooking it
to feed a party of French-speaking Swiss
at the Hotel Calesa Real.
The hornets that failed to return
stain the fountain's edge,
the waters washing and washing away at them,
continuing throughout the night
their whisperings of ablution
where no one stirs,
to the shut flowerheads and the profuse stars.

Oaxaca

IN THE EMPEROR'S GARDEN

It began
With Monsieur de la Borde in a white wig.
His aim was to make this garden in the south
    Perform in miniature with the same
Blithe, musical display as does Versailles
    Or (say) the Luxembourg. But Mexican at heart,
Although it started with this clear intention,
    The urgent earth at once deterred it,
The situation – a steep hillside –, the flora
Of tropical exuberance. The masons tried
With walls, summerhouses and cascades
    To abide by that intention, but their touch
So lay on each pond and fountain you are forced to say
    One thing: This is not French. It hangs
Precipitously above a deep ravine
    When it should have been level. The mangoes,
The sapotes and the Indian laurel, all
    Outgrew the proportions of the place,
Interlaced and roofed out the sky
    With an opaque and sombre canopy:
They are not, as they were intended to be,
    Decorative features of the garden: they
Are the garden itself where one cannot see
    The trees for the forest. Ducks and geese
Lead a lonely, dignified existence here,
    On waters beneath impermeable shade,
Surrounded by arbours of rose and jasmine
    And fountains that for a century have not played.
Hispanicized José de la Borda
    Was long since dead when the Emperor arrived.
The country already slipping out of his hands,
    He planted gardens and bandstands in every town
And there they took. He dreamed
    While the opposition massed and the army fled,
Of extending his empire even now,
    And the trees nodded their absent-minded accord
Above the borders of el Jardin de la Borda.
    A tangle of coffee trees replaced the beds of flowers
Where the Emperor and his Empress strayed alone
    In the company of waters, leaves and stone,

As the retaining walls bequeathed their symmetry
  To the volcanic, shifting soil, their masonry
Silently easing its spoils into the abyss beneath.

The garden at Cuernavaca was a favourite haunt of the Emperor
Maximilian (1832–67). Installed and then deserted by Napoleon III,
Maximilian was shot by the republican opposition.

HUDSON RIVER SCHOOL

for Dorothy and Bill

We drove to the river to see if the shad were rising
Down vistas the painters once flooded with Claudian gold:
The uncompleted spring was still dividing
Shadow-pied acres between the sun and cloud.

As the heat flushed through, you could feel the power of summer
Waiting to overtake the spring only half-begun:
'Where are the fishers?' we asked a woman there,
'You'll see them,' she said, 'once the shad are beginning to run.'

They were not running today here, and the lines
Casting infrequently out from coverts along the shore,
Were merely the first and desultory signs
Of what must turn fever and fervour as more and more

Shad thronged the channels – shad that once filled the fords
In such abundance, you could not ride through
Without treading on them. They have outlasted the herds
Of buffalo, the pigeon flocks that even the painters knew

Only by hearsay, for their retrospective gold
Was painting a loss already. We turned to climb
Out of the valley: Van Winkle's mountains showed
Heaped massively up along the horizon line:

As the hills rose around us we began to see –
Only a few at first, then flare on flare
The blossoms that blew from the shadbush, tree on tree,
Whitening the crests in the currents of the air:

The shadblow that comes when the fish are coming
– Spring brings a yearly proof the legend is true –
Told plainly that plenitude could not be long
In reaching the valley to foison the river anew.

ALONG THE MOHAWK

The child floats out his Indian cry
Across the river. Where did the Mohawks go to?
It arrives with the breath of mudflats as we pass by
Down the disused railroad's asphalted avenue.

What is it the fishers who crouch at the river bends
Hope to hook from this discoloured water?
A rabbit, scenting our approach, pretends
As it freezes in the brush, to exist no more.

You can hear the swish of the cars beyond
The far-bank trees: the even wavelets sway
Up to muddy margins: a docile pond
The river appears, no longer a waterway –

Way of the Algonquin as they came down,
Silently scaled the unmanned stockade,
Dropped into the streets and razed the town,
And the cries of victor and victim carried

Up the river reaches and along these shores
To where a rabbit crouches against invaders
And a railroad, silenced by the swish of cars,
Is the pathway fishers follow across this glade.

PINES

The pines are the colour of the day. They take in
  From the changes of sky and from the bay
Each prevailing, passing shade – thus blue
  Becomes the colour of trees – as true as
And no more to be gainsaid
  Than that hint of red the light is proffering
Between the trunks, only to deny it:
  Sun now burns the whole forest black
With shadow, up to the very crow's nest
  Of each masted pine-tree: there
Such birds as range and turn in the setting light
  Rise like the spirits of trees estranged
That have yielded themselves to a single certain shade
  And, as if by a welling from within,
Darken towards the darkness night has made.

IN THE STEPS OF EMILY CARR

When she got there
The place had disappeared: only a line
Of totems straggled along the bayside
Tipped from the true by winds that patterned through
The breast-high grass the ruins sat in.
She began putting the place together
In careful paint – and then the cats,
An army of them out of every quarter
Of the dank, forsaken clearing, crept
Closer and closer in, yellow-eyed and lean,
Purring, pleading to be taken back
Into the circle of recognitions they had known
Before the dominion of the nettle and the rank
Ferocity of sea-grass, bushes overgrown
And forest dense behind them with the silence
That inhabited it. She left them
Doleful at the very water's edge,
The sole attendants of the wooden goddess
Who still stood, waist-deep in strenuous growth
Amid the odours of rot and humus. They had tasted
The ineradicable sweetness of a human good.

Emily Carr, Canadian painter
(1871–1945)

NORTH WITH LAWREN HARRIS

Here is the glacier of no return. Ice
   Opens into fields and furrows, dry
And hard, and scarcely with a fissure in it –
   A single causeway across silence,
Paved from side to side with white,
   A marble thoroughfare between the tombs
Of a city buried and set round
   With cliffs, that seem as insubstantial
As the blue that bounds them. The scene
   Looms so removed not merely from the presences of men
But from their thoughts, so treelessly complete
   It seems a world no man has known.
Even the spirits that congealed these slopes
   Have perished long since here –
White archangels that heaped them up for tombs,
   Stretched in their shrouds of brilliant hoar.
Too many winters tempered what you drew and saw.

Lawren Harris, Canadian painter
(1885–1970)

FAR POINT

The road ends here. If your way
lies north, then you must take
to the forest or the bay. A café
which is a poolroom which is a bar
serves clams and beer;
the woman who brings them in,
a cheerful exile here,
counts out coins
'Eins, zwei, drei . . .'
with the queen's head on them.
'Look!' I hear her say: a skiff
with an outboard goes past the window.
It's from the island (a strip of sand
with pines and houses on it) and a deer
is swimming in its wake. 'It belongs
to the people in the boat. They should mark it.
I knew a couple who tamed a seal.
It would swim behind *them*, too,
then one day somebody shot it.'
'How would you mark a seal?' I say . . .
'It's easy to mark a deer.
You know how he found it?
It was being born. Out in the woods.
He couldn't resist touching the fur
so the mother abandoned it –
it was the smell of a man on it frightened her.'
The man opens the door:
a group of Indians are playing pool,
the usual clientele on the bar stools.
They look like lumberjacks. Americans,
they came north to stay out of the army
and never went back. Now they are grey.
He joins them and the deer
paces the veranda and tries
with its great deer's eyes
to look through
the deceptions of the long window
and find where he is.
The pool-players, backs to the light,
stand facing away
from centuries of clambake, potlatch

and tribal ferocities
down a totem-guarded coast.
The poles rotted and the seeds,
dropping inside their crevices,
turned them back into trees.
The cue's click, click
rehearses its softly merciless music
ticking away the increment
of unwanted time. We return
to the car, past fragile shacks
whose cracked white paint
the sea air is picking apart.
We discover the deer once more
that gazes right through us,
then catching sight of a pair of dogs
arcs off to play with them,
perhaps thinks it's a dog.
Across the blue-grey strait,
the ragged ideograms of firs
in a rising and falling fog:
clouds are what we appear to contemplate
above them, then the mist
stirs, sails off
and we see it is summits
we are peering at,
that go on unveiling themselves
as if they were being created.

ALGOMA

(Northern Ontario)

Each day advanced the passing of the leaves:
The maples first – so much of richness there,
Colour could not have held beyond that pitch,
Yet fallen, frosted, when October sun
Sets them alight once more, their thousands burn
At first intensity. The yellow leaves
Followed the scarlet fast and left all bare
Except the spruces whose interior dark
Densens their louring greens against the light.
The Dipper measures out sky spaces now,
Capella, unencumbered, pricks the eye
Where leaves on leaves had stalled it night on night,
And the first flakes thickening down the wind
Solidify the dark, then hide the sky:
Nobody owns these lands with only the pole behind.

LINES WRITTEN IN THE BAY
OF LERICI

These giant butterflies feel out the wind
   And circle one another, swaying so safely round
No craft is spilt or split: they lift
   Their wingfuls of air and lifted by it
Bear their own burdens lightly, true
   To the coast curves as the hulls pursue
This swerving waltz of wake lines:
   The salient points, the salient sails
Reassembling, finding new spaces for themselves,
   The sea is a blue page to the signs
They write out on it. Like lizard tails
   (In shape, not shade – in rhyme, not reason),
The veering of these trails! – and deft as lizards
   That lighten across the walls of this high garden,
Where our feet on the sloping ground of a terraced hill,
   Teach us to tell the pitch and scope
Of a score of sails, tacking and steadying
   There below on a bay cupped-in
From the full breath of the September Apennine
   That autumn is whitening, wintering already.

THE LABYRINTH

for Astrid

Generations labyrinthed this slope – wall
   On low wall, then pergolas of vines
To roof them in. Two workmen
   Pensively complete a further parapet
And they, as we wind up past,
   Give us good day. At our return we find
One of the pair a level higher.
   He greets us yet once more and vanishes –
Only to reappear below us carrying with care
   The single brick that he was looking for,
Exiting by a sidepath to press home
   The one piece missing that (once mortared-in)
Will answer to the picture in his mind
   Perhaps, until all fits –
A provisional offering to the god of limits.

CARRARA REVISITED

Only in flight could you gather at a glance
So much of space and depth as from this height;
Yet flight would blur the unbroken separation
Of fragile sounds from solid soundlessness –
The chime of metal against distant stone,
The crumple and the crumble of devastation
Those quarries filter up at us. Our steps,
As they echo on this marble mountain,
Make us seem gods whom that activity
Teems to placate. But not for long. The hawk
Stretched on the air is more a god than we,
And sees us from above as our eyes see
The minute and marble-heavy trucks that sway
Slowly across the sheernesses beneath us, bend on bend,
Specks on an endlessly descending causeway.

THE HOUSE IN THE QUARRY

What is it doing there, this house in the quarry?
  On the scrap of a height it stands its ground:
The cut-away cliffs rise round it
  And the dust lies heavy along its sills.
Still lived in? It must be, with the care
  They have taken to train its vine
Whose dusty pergola keeps back the blaze
  From a square of garden. Can it be melons
They are growing, a table someone has set out there
  As though, come evening, you might even sit at it
Drinking wine? What dusty grapes
  Will those writhen vine-stocks show for the rain
To cleanse in autumn? And will they taste then
  Of the lime-dust of this towering waste,
Or have transmuted it to some sweetness unforeseen
  That original cleanliness could never reach
Rounding to insipidity? All things
  Seem possible in this unreal light –
The poem still to be quarried here,
  The house itself lit up to repossess
Its stolen site, as the evening matches
  Quiet to the slowly receding thunder of the last
Of the lorries trundling the unshaped marble down
    and past.

AT THE AUTUMN EQUINOX

for Giuseppe Conte

Wild boars come down by night
  Sweet-toothed to squander a harvest
In the vines, tearing apart
  The careful terraces whose clinging twines
Thicken out to trunks and seem
  To hold up the pergolas they embrace.
Make fast the gate. Under a late moon
  That left the whole scene wild and clear,
I came on twenty beasts, uprooting, browsing
  Here these ledges let into the hillside.
They had undone and taken back again
  Into their nomad scavengers' domain
All we had shaped for use, and laid it waste
  In a night's carouse. Which story is true?
Those who are not hunters say that hunters brought
  The beasts to this place, to multiply for sport
And that they bred here, spread. Or should one credit
  The tale told of that legendary winter
A century since, which drove them in starving bands
  Out from the frozen heartlands of the north?
Ice had scabbed every plane and pine,
  Tubers and roots lay slabbed beneath the ground
That nothing alive or growing showed above
  To give promise of subsistence. They drove on still
Until they found thickets greening up through snow
  And ate the frozen berries from them. Then
Down to the lowland orchards and the fields
  Where crops rooted and ripened. Or should one
Go back to beginnings and to when
  No men had terraced out these slopes? Trees
Taller than the oaks infested then
  These rocks now barren, their lianas
Reaching to the shore – the shore whose miles
  On miles of sand saw the first approach
As swarms swam inland from the isles beyond
  And took possession. Are these
The remnant of that horde, forsaking forests
  And scenting the orchards in their wake? I could hear them
Crunch and crush a whole harvest
  From the vines while the moon looked on.

A mouse can ride on a boar's back,
  Nest in its fur, gnaw through the hide and fat
And not disturb it, so obtuse is their sense of touch –
  But not of sight or smell. I stood
Downwind and waited. It takes five dogs
  To hunt a boar. I had no gun
Nor, come to that, the art to use one:
  I was man alone: I had no need
Of legends to assure me how strange they were –
  A sufficiency of fear confessed their otherness.
Stay still I heard the heartbeats say:
  I could see all too clear
In the hallucinatory moonlight what was there.
  Day led them on. Next morning found
These foragers on ground less certain
  Than dug soil or the gravel-beds
Of dried-up torrents. Asphalt
  Confused their travelling itch, bemused
And drew them towards the human outskirts.
  They clattered across its too-smooth surfaces –
Too smooth, yet too hard for those snouts
  To root at, or tusks to tear out
The rootage under it. Its colour and its smell,
  The too-sharp sunlight, the too-tepid air
Stupified the entire band: water
  That they could swim, snow that had buried
All sustenance from them, worried them far less
  Than this man-made ribbon luring them on
Helpless into the shadow of habitation.
  The first building at the entrance to the valley
Had *Carabinieri* written across its wall:
  Challenged, the machine-gunned law
Saw to it with one raking volley
  And brought the procession to the ground,
Then sprayed it again, to put beyond all doubt
  That this twitching confusion was mostly dead
And that the survivors should not break out
  Tusked and purposeful to defend themselves.
Blood on the road. A crowd, curious
  To view the end of this casual hecatomb
And lingeringly inspect what a bullet can do.
  It was like the conclusion of all battles.
Who was to be pitied and who praised?
  Above the voices, the air hung

31

Silent, cleared, by the shots, of birdsong
    And as torn into, it seemed, as the flesh below.
Quietly now, at the edges of the crowd,
    Hunters looked the disdain they felt
For so unclean a finish, and admired
    The form those backs, subdued, still have,
Lithe as the undulation of a wave. The enemy
    They had seen eviscerate a dog with a single blow
Brought into the thoughts of these hunters now
    Only their poachers' bitterness at flesh foregone
As their impatience waited to seize on the open season,
    The autumn equinox reddening through the trees.

OXEN

There are no oxen now in Tuscany.
Once, from any hill-top, you might see
The teams out ploughing, the tilled fields stretch away
Wherever those bowed heads had chamfered round
The swelling contour. The first I ever saw
Strode over ground in such good heart their ease
Was the measure of its tilth. The ploughman knew
His place – behind his beasts, and at the head
Of all the centuries that shaped these hills.
Once, *Belle bestie*, I murmured to myself,
Passing a stall that opened on the road
And catching sight of oxen couched inside.
Belle bestie a voice replied, and I
Was ushered in to touch and to admire
The satin flanks, the presence on the straw.
I recognized the smell, recalled the warmth
Of beasts that rustled all one winter night
In the next room: the stars across Romagna
Pinned out the blackness of the freezing sky
Above a plain that sweat of ox and man
Had brought into fruition. Once I saw
An ox slain in an abattoir. The blood
That flushed the floor was dark and copious –
More than enough to hold the gods at bay
Or bring the dead to speech. The dead spoke then
As every deft stroke of the butcher's men
Revealed an art that was not of a day.
The toughest ground that ever oxen broke
Was by Sant' Antimo. I watched the dust
Turn their slaverings brown and choke the man
Who jolted in their wake and cursed the stones
That cropped out everywhere, unslakeable
The thirst that parched him. He would have been the first
To welcome in the aid that brought his end,
These cheerful tractors turning up the land
Across all Tuscany. No bond of sweat
Cements them to his generations: careless blades
Advancing to the horizon as the clods yield,
Bury both beast and man in one wide field.

THE BUTTERFLIES

They cover the tree and twitch their coloured capes,
    On thin legs, stalking delicately across
The blossoms breathing nectar at them;
    Hang upside-down like bats,
Like wobbling fans, stepping, tipping,
    Tipsily absorbed in what they seek and suck.
There is a bark-like darkness
    Of patterned wrinklings as though of wood
As wings shut against each other.
    Folded upon itself, a black
Cut-out has quit the dance;
    One opens, closes from splendour into drab,
Intent antennae preceding its advance
    Over a floor of flowers. Their skeletons
Are all outside – fine nervures
    Tracing the fourfold wings like leaves;
Their mouths are for biting with – they breathe
    Through stigmata that only a lens can reach:
The faceted eyes, a multiplying glass
    Whose intricacies only a glass can teach,
See us as shadows if they see at all.
    It is the beauty of wings that reconciles us
To these spindles, angles, these inhuman heads
    Dipping and dipping as they sip.
The dancer's tread, the turn, the pirouette
    Come of a choreography not ours,
Velvets shaken out over flowers on flowers
    That under a thousand (can they be felt as) feet
Dreamlessly nod in vegetative sleep.

CHRONOCHROMIE

The thrushes are singing their morning plainchant,
   Colouring time – rehearsing once again
In flood and droplet, cascade and chime
   Those centuries before our coming here
To measure and to minute out
   All to our purposes. Now we go
Back with the birds to before we were.
   We cannot stay long. Only the angels
Could listen out that song
   Through the millennia of our silence. Yet angels
And other feathered things have in common
   Merely their wings, for thrushes
Both sing and stab. And we alone
   Who invented angels, but not the birds,
Hear brink and beginning in their wordless words –
   Hear space begetting time once more
In the falls and flourish of that coloratura.

RUSKIN REMEMBERED

What is it tunes a'Scottish stream so fine?
Concurrence of the rock and of the rain.
Much rain must fall, and yet not of a sort
That tears the hills down, carries them off in sport.
The rocks must break irregularly, jagged –
Our Yorkshire shales, carpenter-like, form merely
Tables and shelves for rain to drip and leap
Down from; the rocks of Cumberland and Wales
Are of too bold a cut and so keep back
Those chords their streams should multiply and mingle.
But there must be hard pebbles too – within
The loosely breaking rock, to strew a shingle
Along the level shore – white, for the brown
Water in rippling threads to wander through
In amber gradations to the brink, the ear
Filled with the link on link of travelling sound,
Like heard divisions, crisp above a ground,
Defining a contentment that suffices –
As walking to unblent music, such as this.

A RUSKINIAN FABLE RETOLD:
COURTESY

– or however you'd denote
the behaviour of this quartz,
living side by side
with the seemingly more modest
green and slender
mineral called 'epidote':
they can't go on
growing together now
much longer: the quartz
five times as thick
is twenty times the stronger.
Sensing, at the very crown
and self-built summit
of its own existence,
the presence of its weak
persistent neighbour,
it pauses there and lets
the pale-green film
of epidote grow past
to occupy the space
beyond it. The cost
of this well-bred hesitation?
Its own crystal life.
No: 'courtesy'
is not the word: courtesy
is closer to common sense
than immolation.

BLUEBELLS

Bluebells! we say seeing the purple tide
   Overflow from the wood to meet us.
If we could fly above them, we could read
   The sprawled, imperial hieroglyph of this spread
Above whatever nutriment of earth has fed
   Their fresh advance. The scent
Drenches the summer air it cools
   With the fulness of their presence: they
Swarm down the woodbank, a flower army.
   If the angelic orders were visible in time,
Then these might glow as the iridescent shadow
   Of such splendour. If they are blue –
Silent the bell-mouths crowding on each stem –
   It is only our words so call and colour them.

HARVEST

for Paula and Fred

After the hay was baled and stacked in henges,
We walked through the circles in the moonlit field:
The moon was hidden from us by the ranges
Of hills that enclosed the meadows hay had filled.

But its light lay one suffusing undertone
That drew out the day and changed the pace of time:
It slowed to the pulse of our passing feet upon
Gleanings the baler had left on the ground to rhyme

With the colour of the silhouettes that arose,
Dark like the guardians of a frontier strayed across,
Into this in-between of time composed –
Sentries of Avalon, these megaliths of grass.

Yet it was time that brought us to this place,
Time that had ripened the grasses harvested here:
Time will tell us tomorrow that we paced
Last night in a field that is no longer there.

And yet it was. And time, the literalist,
The sense and the scent of it woven in time's changes,
Cannot put by that sweetness, that persistence
After the hay was baled and stacked in henges.

THE GARDEN

for the same

                                      And now they say
Gardens are merely the expression of a class
Masterful enough to enamel away
    All signs of the labour that produced them.
This crass reading forgets that imagination
    Outgoes itself, outgrows aim
And origin; forgets that art
    Does not offer the sweat of parturition
As proof of its sincerity. The guide-book, too,
    Dislikes this garden we are descending through
On a wet day in Gloucestershire. It speaks
    Accurately enough of windings and of water,
Half-lost pavilion, mossy cascade,
    But is afraid 'the style is thin.' One must smile
At the irritability of critics, who
    Impotent to produce, secrete over what they see
Their dislike or semi-assent, then blame
    The thing they have tamed for being tame.
But today, see only how
    Laden in leaves, the branchwork canopy –
Bough on bough, rearing a dense
    Mobile architecture – shudders beneath its finery
In cool July. Heat, no doubt,
    Would flesh out the secret of this garden where
(Or so it's said) the man who imagined it
    Could wind down to find
His gypsy inamorata waiting there
    By the hidden lake.
                             There are three lakes here
And a fluttering curtain of rain that falls
    Differently in each. The first
Lies open to the farmland and it takes
    The full gust and disordering of the weather
Across its surface. The second –
    We have descended further now
Bending our way in under each low bough –
    Shelters between the hills' high shoulders,
And so the green, smooth plain of water
    Lies taut under the nail-points of the rain.

We must enter next a key-hole door
Into darkness: through a rough window-slit
We catch a runnel wrinkling over stone,
And the pool that stretches to receive it
So fills the aperture we can not take in
Its true extent: we are all eye –
Which is not eye enough to outdo
The dark we are trying to gaze out through.
A twist in the tunnel: light! We are delivered
And now we can freely move
Beneath a pergola 'in the precise arch'
(I quote to show the disapproval I disapprove)
'Of a railway terminus.' This is no end
However, but the start of the final lake.
You can see the rain withdraw across
This widest of the waters, the transparent scrim
Suddenly towed aside, and calm
Flowing up to its receding hem:
Fish in the cloudy depths might well be swimming
Through sky such as threatens us still.
We have the hill to re-ascend, and do,
Up to the formal garden at its summit:
The statuary, the espaliered avenue
Ignore the twisting path. The descent
To the hidden lake now hoards from sight
The walks and walls, the subterranean door
Into an imaginary place that time
Turned real. The imagination hovers here
Half rebuked, with its Doric and Chinese;
Nor can a planned secretiveness outdo
The cool green of that chamber
Shut from view beneath the gloom
Of the copper beech. Its tent conceals
Not darkness, but an inner room,
An emerald cell of leaves whose light
Seems self-sustaining, and its floor
A ground for the reconciling of our dreams
With what is there.
                        So here we stand,
We two, and two from another land,
To meditate the gift we did not ask –
The work of seasons and of hands unseen
Tempering time. What has not disappeared

Is a design that grew – ultimately to include
    (Beside plants of oriental and American species)
Us four, in its playful image of infinity,
    The whole of it assembled with a view
To generations beyond the planter's: there is nothing here
    We shall ever own, nothing that he owns now,
In those reflections of summer trees on water,
    This composure awaiting the rain and snow.

LETTER TO UEHATA

Since I returned, the trees have a Japanese look,
   And bare in their wintry sinuousness seem
To retrace in air the windings of those paths
   That followed so faithfully the swelling ground,
Then lost for an instant, came back into view
   With the trees they wound on through, reflected
(Borrowed, as you would say) by some pool.
   That landscape was arranged – to reflect
And reflect again the grain and grandeur
   Of the world we see, and that the centuries
Have unrolled to now, as if time
   Were itself the paradigm of a path
That has brought me to where I can read
   In the bareness of the trees a double scene –
Where I am now and where we both were then.

APPLES PAINTED

for Olivia

He presses the brush-tip. What he wants
  Is weight such as the blind might feel
Cupping these roundnesses. The ooze
  Takes a shapely turn as thought
Steadies it into touch – touch
  That is the mind moving, enlightened carnality.
He must find them out anew, the shapes
  And the spaces in between them – all that dropped from view
As the bitten apple staled on unseen.
  All this he must do with a brush? All this
With a brush, a touch, a thought –
  Till the time-filled forms are ripening in their places,
And he sees the painted fruit still loading the tree,
  And the gate stands open in complicity at his return
To a garden beneath the apple boughs' tremulous sway.

WINTER

There is no light left. And yet
  A glow light covers and colours
The ground beneath the trees: it is leaves
  Have fallen from the beech – the same red
As the earth that shows between them, smouldering
  On shining trunks, a fire
The mirror-surfaces of rain redouble
  In unlit air. Mirror on mirror
Hands it on into the wood, raises
  The light of this ground-level brazier
Half way up the boles and into our eyes
  In a punctual sunset, where we had looked
For darkness underneath the night-wide
  Moss-fattening down-drift of the wet.

THE CYCLE

So fine the snow
You must look to the blackness of the trees to show
The shaping of its grain against the wind.
The sky is as white as that descent,
The ground, too, hides what it receives
Transforming to a blent and cleanlier white
The hanging threat vouching for more and more.
The house is going under as the drift
Climbs walls and door and, sieving through,
Gathers finely in a threshold line.
It appears to mean that we – the ones who've seen
A flood seeping across that limit –
Stand guarding a frontier, the incursion
Refusing to disband until it must, and then
Grudgingly reveal the green the spring
Thrusts back at us, the cycle rebegun.

TWO POEMS FOR FAY GODWIN

I LEAPING LURCHER

More flag than dog –
more pennant than drapeau –
see it stretch mid-air
to clear the topmost
barbs of a fence-wire,
and hang for the infinitesimal
flick a shutter takes
to imprison it and show
(what did not quite occur)
the poised, heraldic flight
caught flat as a weathervane
to be unwrought by time,
metal to muscle, forefeet
prepared already
to take the fall
they must prevent
and steady into movement

II ROYAL MILITARY CANAL

A phalanx of sheep,
the sun behind them,
defended the further shore:
shadows sloping towards us,
their heads and shoulders
cast across grass
a battlement in silhouette
that ceased at the water: the real
heads (a wholesale decapitation)
overshot this shadowy parapet to reappear
in reflection, upside-down and yet
as alert to our intentions
as those heads-on-shoulders
awaiting our first move:
I opened both eyes,
clapped my hands
and the snapshot
fell apart bleating.

47

FELIX RANDAL

Or the Black Economy

The farrier comes with his forge. He shoes
  For cash only. The ash of a portable fire
Leaves no trace in the account books of the nation,
  The acridity of singeing hooves prompts no enquiry
And the sounds of the hammer are swallowed by the air
  As scavenging farm dogs gobble the parings.
The collapsible forge ballasts his swaying van:
  Unpursued by conscience or by priest – Felix is gone.

MUSIC AND THE POET'S CAT

Your ears resented the discords,
the scream of the woodwind
in *Tapiola* before the whirlwind
spreads through the orchestra
diminishing those centuries
of chorales and firm persuasions.
Now that you are gone,
run into marl and waterdrops
beneath your stone,
once more I listen
and my ears take in
those selfsame notes
across this rent
in time and I
resent both it and them.

ODE TO DMITRI SHOSTAKOVICH

I

To what far room of never-to-return
The raw brass singles out and calls us –
A tale too often told, one old already
Long before the pen of Mandelstam
Entangled itself in the Georgian's moustaches.
Those feelers found him out, and you survived
To play the fool and to applaud the play
That you must act in. Notes told less than words
And now tell more, each vast adagio dense
With the private meaning of its public sorrow.

II

You stole the Fate Motif from Wagner's *Ring*
(Great artists steal and minor merely borrow) –
Fate had declared itself as daily fact:
This day might be the allotted span. Tomorrow . . . ?
Stalin was dead. But not his heirs, and not
The memory imprinted in the nerves, the heart.
Light-eared, light-fingered, you had earned your share
Of that absurd prestissimo from *William Tell*,
To accompany an endlessly running man
In one of the silent comedies of nightmare.

III

Inscribing score on score with your motto theme –
Mnemonic of survival, notes for a name –
'I am still here,' you signal, yet once more,
And now that you no longer are, the same
Chime recurring takes on the whole of time
Out of a permanence few of us can have.
In the photograph you pass smiling to the grave.

MOONRISE

I cannot tell you the history of the moon,
Nor what they found contained within its dust.
All I can say, this December afternoon,
Is that it rises early as the last
Of the crows are spying out a way
In semi-darkness to a darkened nest,
Its phosphor burning back our knowledge to
The sense that we are here, that it is now.
Against the east, the tautness of its bow
Is aiming outwards at futurity
And that will soon arrive, but let it be:
The birds are black on the illuminated sky
And high enough to read the darkness here
By this risen light that is bringing tides to bear.

ORION OVER FARNE

for John Casken

The growling of the constellations, you said –
   A more ferocious music of the spheres
Where, above Farne, the Scorpion tears
   Orion still, teaches him hunt elsewhere,
But hunt he will – and here
   Over the breathing body of the sea,
Heard through the darkness and the star-rimed air
   To the sharp percussions of the tide on scree.
Close to, a poet feeds this frosty soil
   Where the November constellation sets
As storms on storms begin, as the spoils
   Of another year are scattered and constellation
On hunted constellation grinds and growls.

CHANCE

I saw it as driving snow, the spume,
   Then, as the waves hit rock
Foam-motes took off like tiny birds
   Drawn downwind in their thousands
Coiled in its vortices. They settled
   Along ledges and then fell back,
Condensed on the instant at the touch of stone
   And slid off, slicking the rock-sides
As they went. The tide went, too,
   Dragging the clicking pebbles with it
In a cast of chattering dice. What do they tell
   These occurrences, these resemblances that speak
      to you
With no human voice? What they told then
   Was that the energies pouring through space and
      time,
Spun into snow-lace, suspended into flight,
   Had waited on our chance appearance here,
To take their measure, to re-murmur in human sounds
   The nearing roar of this story of far beginnings
As it shapes out and resounds itself along the shore.

THE HEADLAND

A silence lies over the headland like a death
   That has left in the air an echo of the stir
That it has checked – you hear it in the breathing of the sea
   Lipping at the pebbles continuously
Below the cliff, as if it could not articulate
   The word it wanted to deliver, yet bringing to bear
All of the forces it takes to shape one word:
   Unseizably it rehearses an after-life
(The only one certainly there) like that of verse
   That holds its shell to the ear of a living man,
Reminding him that he will be outlasted
   By the scansion in its waves, beating a shore
That is the beginning of the voyage out
   Towards the continuing sunsets, on and on
Cast back across the façades of the shoreline town.

FOR A GODCHILD

Given a godchild,
I must find a god
worthy of her. Dante
refused – in courtesy
(he said) to the god
he venerated, to wipe
a sinner's eyes in hell:
I must tell her of that
one day, and see
that she ponders well
what she takes to be
the dues of deity –
and learn that a god
who harbours anger where
thirst has no slaking,
eyes no ease,
is either of her own
or others' making.

OXFORD POETS

Fleur Adcock
James Berry
Edward Kamau Brathwaite
Joseph Brodsky
Michael Donaghy
D. J. Enright
Roy Fisher
David Gascoyne
David Harsent
Anthony Hecht
Zbigniew Herbert
Thomas Kinsella
Brad Leithauser
Herbert Lomas
Derek Mahon
Medbh McGuckian
James Merrill

John Montague
Peter Porter
Craig Raine
Tom Rawling
Christopher Reid
Stephen Romer
Carole Satyamurti
Peter Scupham
Penelope Shuttle
Louis Simpson
Anne Stevenson
George Szirtes
Grete Tartler
Anthony Thwaite
Charles Tomlinson
Chris Wallace-Crabbe
Hugo Williams

also

Basil Bunting
W. H. Davies
Keith Douglas
Ivor Gurney
Edward Thomas